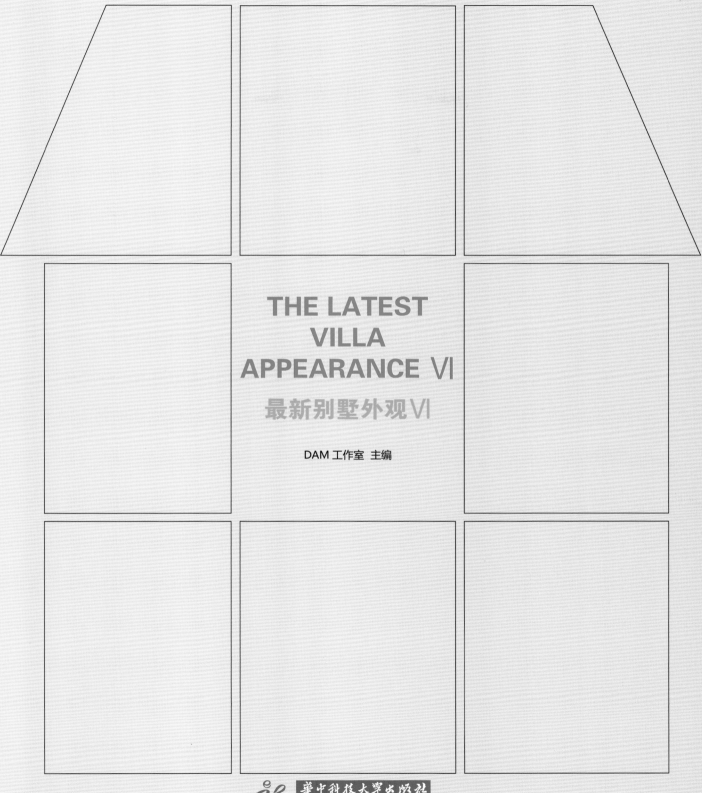

THE LATEST VILLA APPEARANCE Ⅵ

最新别墅外观Ⅵ

DAM 工作室 主编

华中科技大学出版社
http://www.hustp.com
中国·武汉

PREFACE 序言

漫看，一场别墅流光！

"别墅临都门，惊湍激前后"这是唐代诗人李颀较早对别墅的描述。"别墅"一词在我国并非舶来品，而是自古有之，最早称"别业"，又称"别墅"。只是后来在时间长廊里与西方的别墅概念不期而遇，共谱一场别墅流光。

别墅风情，向之往之。于都市的繁华中，看却人世的嘈杂。凡尘之器，倦之厌之。喟叹都市之累，生活之劳。欲归，欲隐，离合之意，皆能近乎？实不尽然。白居易曾有云，"大隐住朝市，小隐入丘樊"，如今，无论朝市，抑或丘樊，皆有不堪之累。只愿选一方山水，留一块滋土，于此处筑一栋楼阁，漫步内心之桃源，聆听自己的怦然心动！

别墅皆人匠，空灵本天成，远观、近赏皆在意境之中、诗话之里。中式别墅的园林之境，日式别墅的禅意深邃，欧式别墅的浪漫与富丽，美式别墅的简约大气，千姿百态，美不胜收。感官之娱，视听之乐，皆可极尽。

目游
漫游于一片葱郁美景，抬眼即拾的美，尽存于别墅宅邸之周。这一片绿色森林，或是一方天然氧吧。有春之青绿，夏之繁茂，秋之凋零，冬之萧瑟，四季之轮回，双目之炫游。其外观，或青砖灰瓦，或红黄炫目，或古石迎面，或全幅玻璃框景，极尽材质与色彩。砖砌的烟囱、整齐的老虎窗、曲面的屋顶、雕花的栏柱……情调，不必刻意寻找；空灵，无需矫揉造作。双目所触，就可算是一场骋怀天地的极目之游。

耳听
慧聪之享，莫过且闭双眼，闲听流水之潺潺。无论是天然水景之自然，抑或是人造水景之灵动，与别墅极致外观配搭，方可尽显水之妩媚与柔和。一片葱绿的盛夏，在这里，更可尽赏蝉之啾鸣，甚至是蜂蝶舞翅那一瞬的震撼，侧耳而听，感叹生命之力量，内心难免不为之倾倒。

呼吸
别墅或许不必拥山揽水，不必有亭台楼阁，但必须有暖暖阳光、满室芬芳。呼吸着暖暖阳光、幽幽花香，才能真正地感受到生命的存在，才能使你的心灵领地独傲于繁琐之尘。于静好岁月里，独坐露台，细品香茗，静听水之律动，闲赏光束之穿梭，呼吸着清风与泥土的清香……如斯般静雅，如斯般温暖，流转时光里的美好，可触、可感。

闲情皆有可寄，逸致亦有可兴，一切美与爱皆有所出。美之缔造者，其心之宽、情之真、艺之妙、思之巧，无不为之折服。感念于建筑之优秀、图片之精美、设计师之慷慨，故在此一并作谢国内外优秀建筑设计师！

目录 CONTENTS

现代风格别墅	Modern-style Villa	004~115
多样风格别墅	Various Style Villa	116~251

Modern-style Villa

现代风格别墅

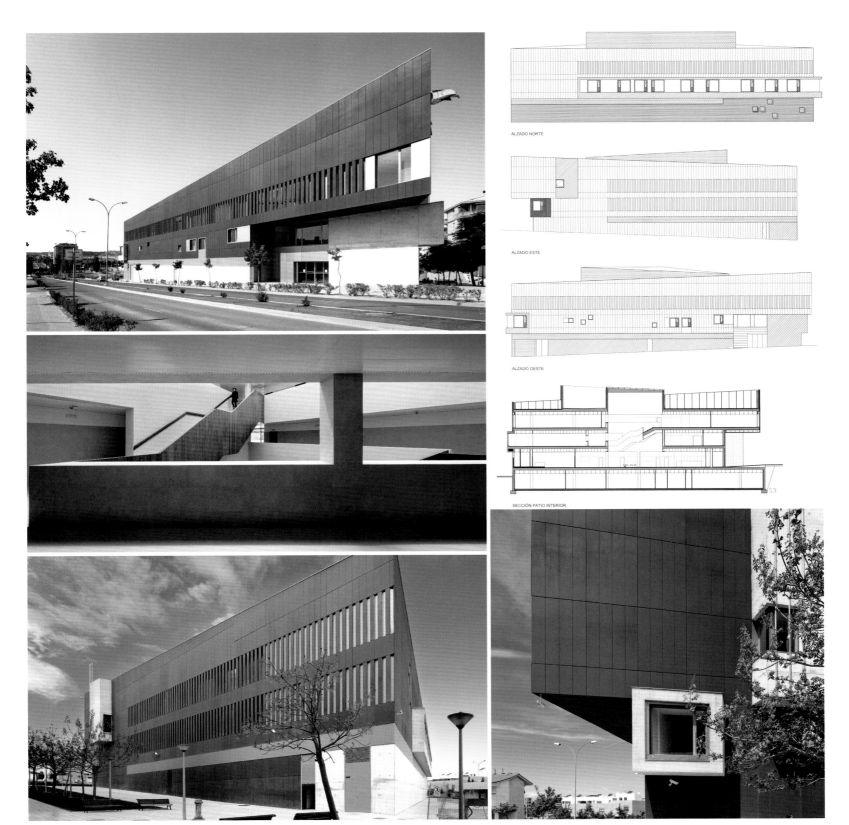

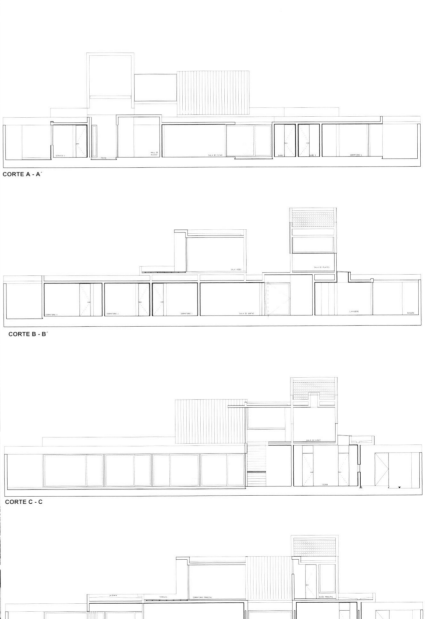

CORTE A - A'

CORTE B - B'

CORTE C - C

CORTE D - D

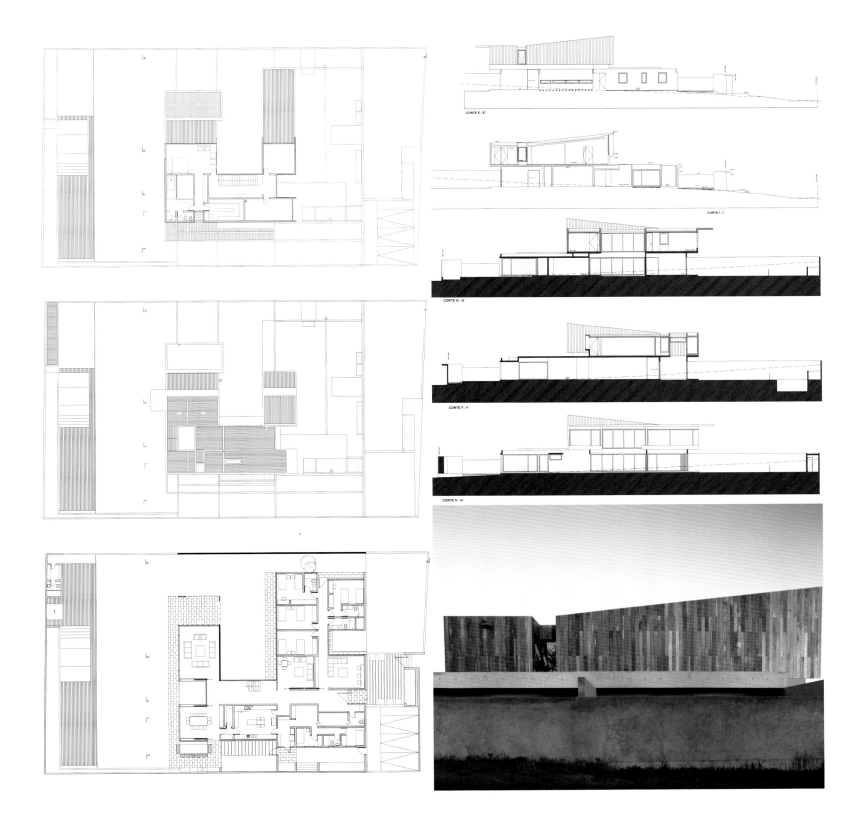

THE LATEST VILLA APPEARANCE VI 011

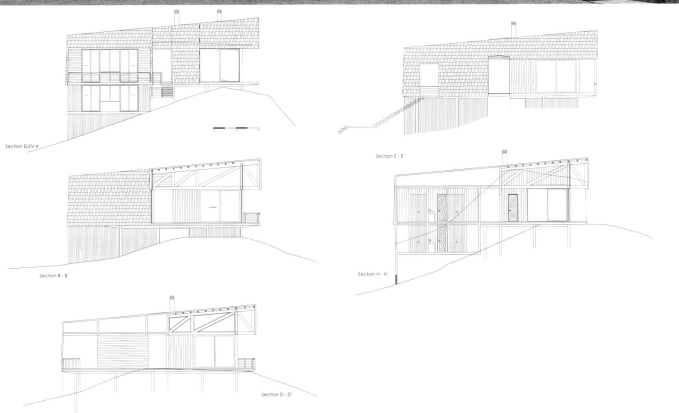

Section ELEV-4

Section E - E´

Section B - B´

Section H - H´

Section D - D´

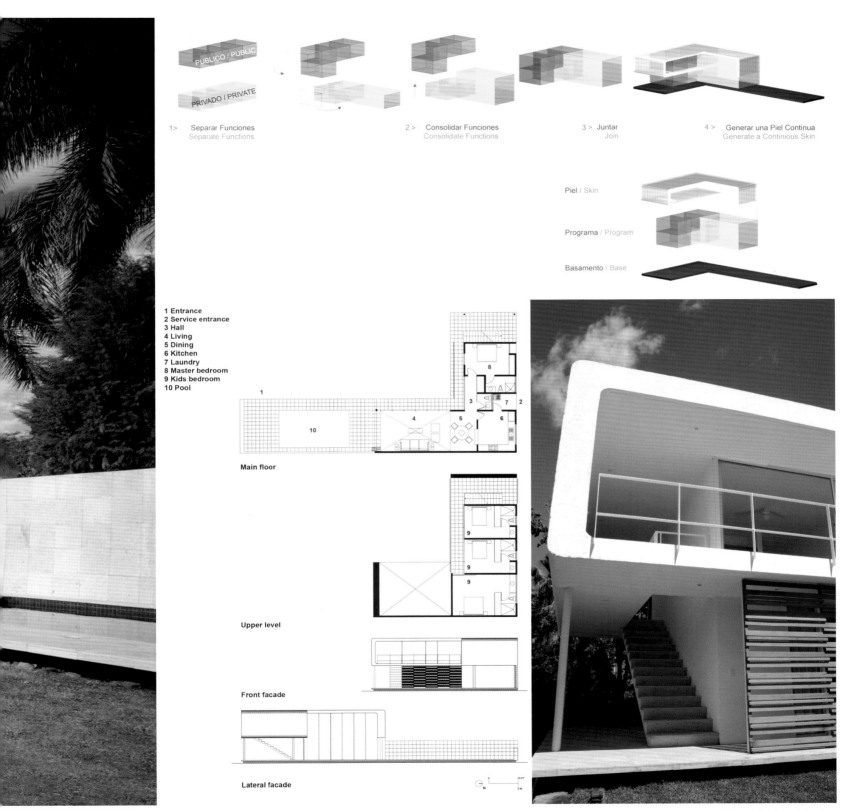

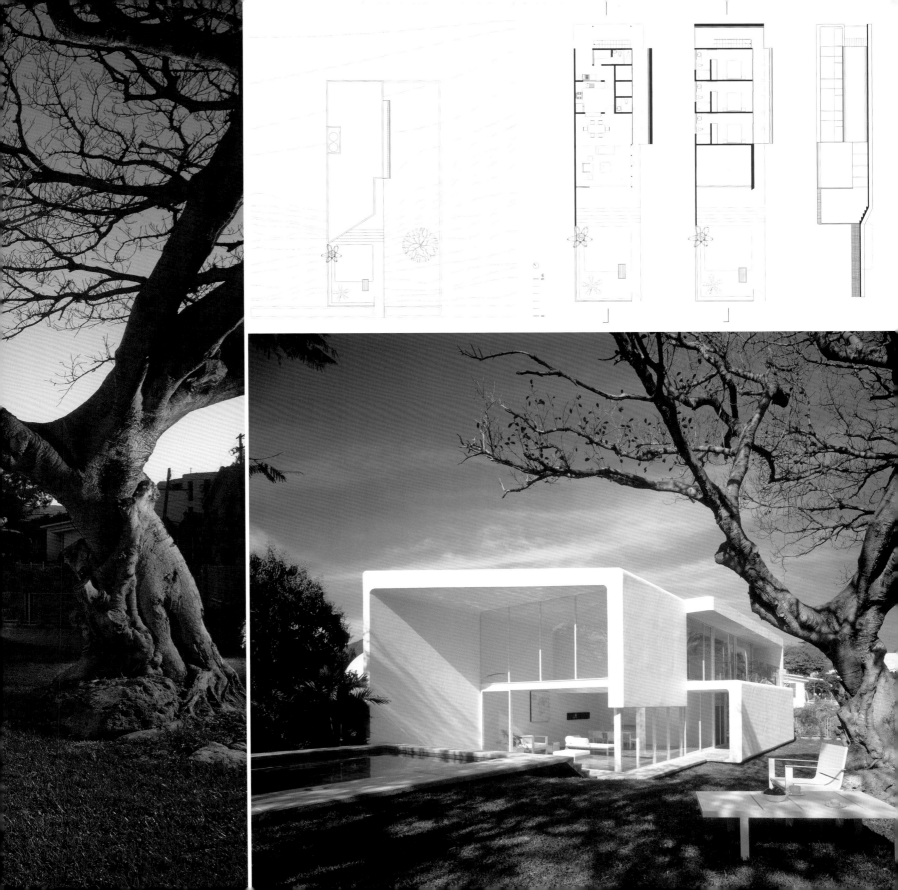

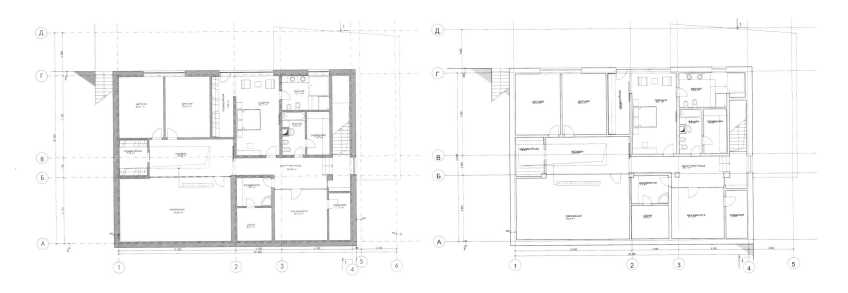

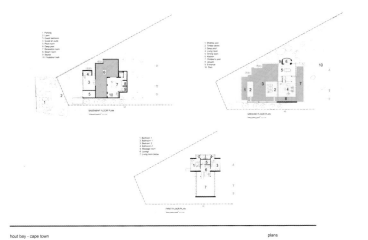

hout bay - cape town · plans

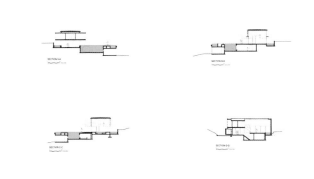

hout bay - cape town · sections

hout bay - cape town · concept

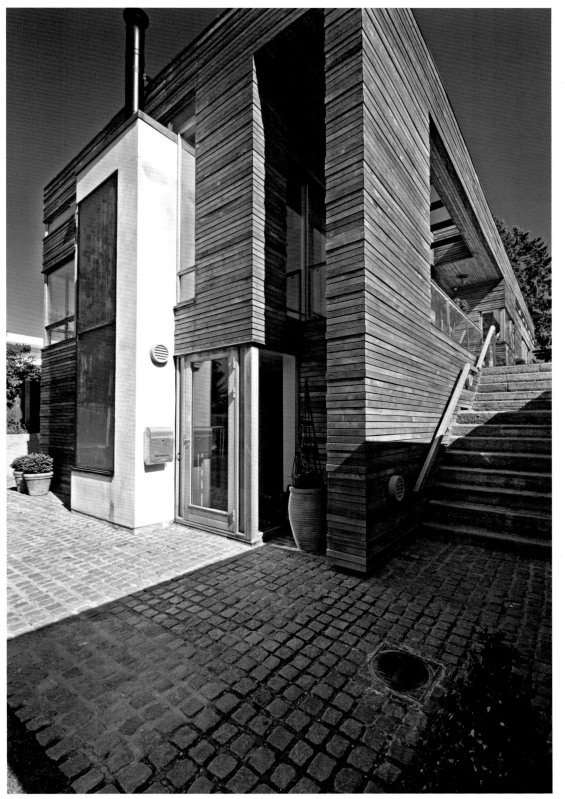

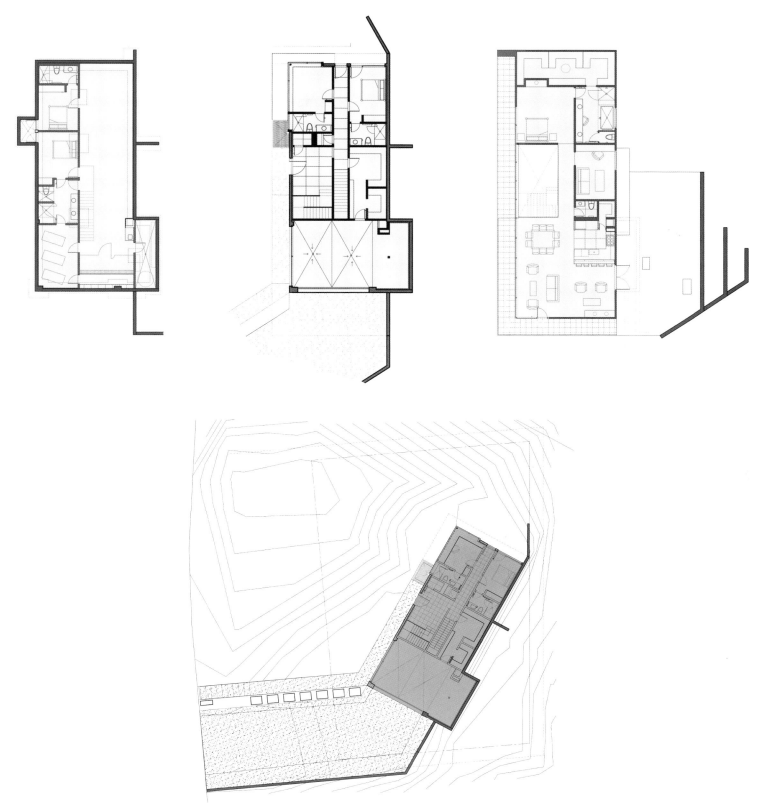

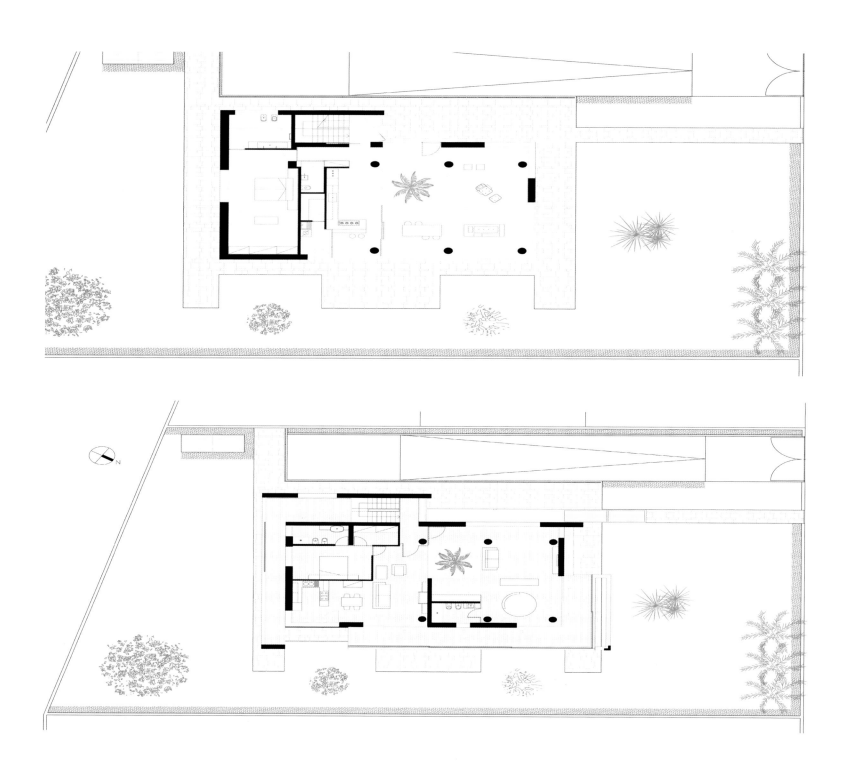

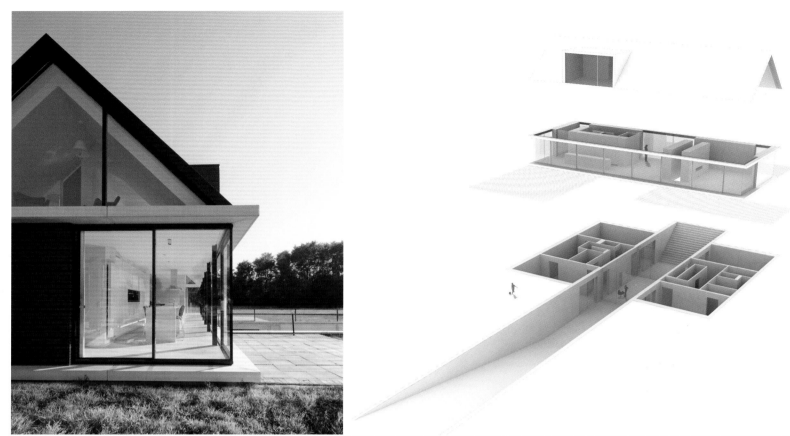

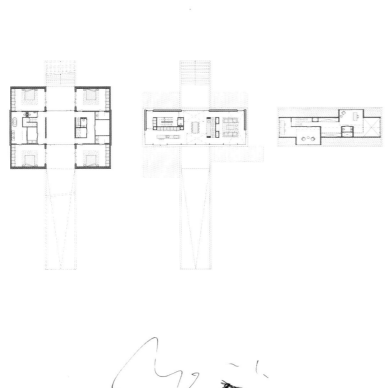
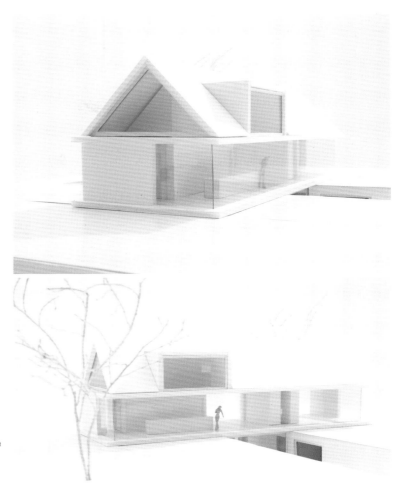

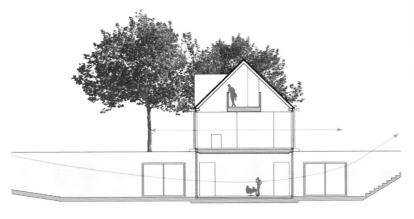

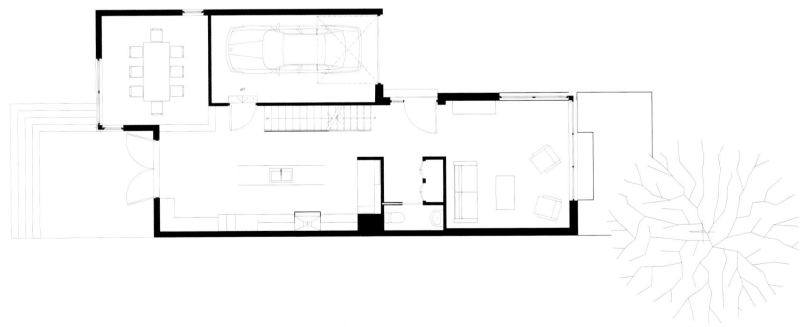

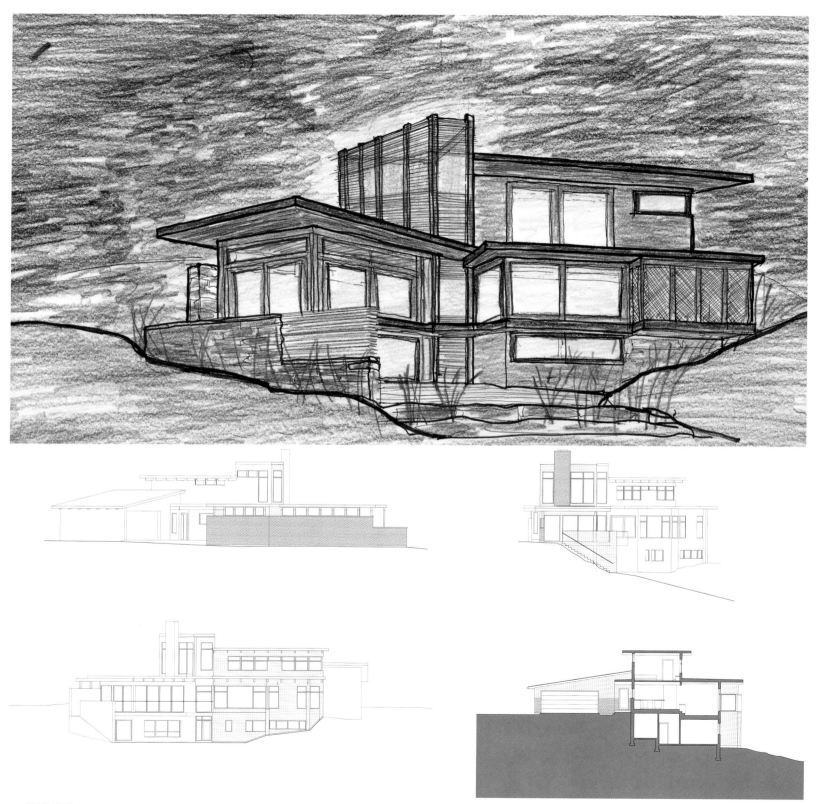

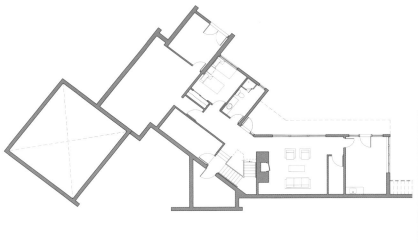

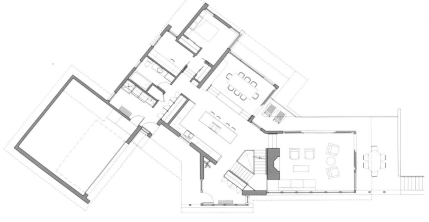

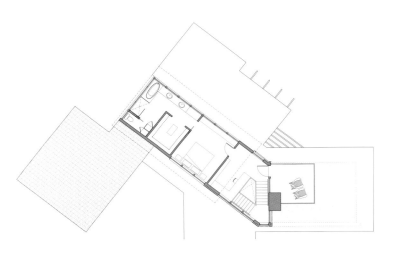

1. Lower Floor
SCALE:3/16" = 1'-0"

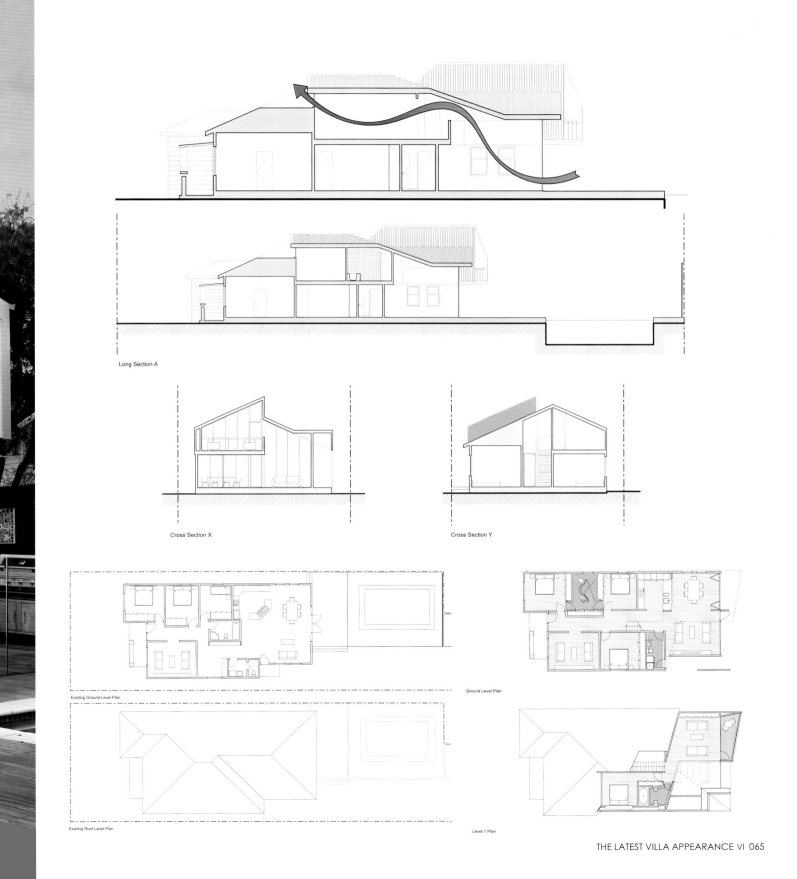

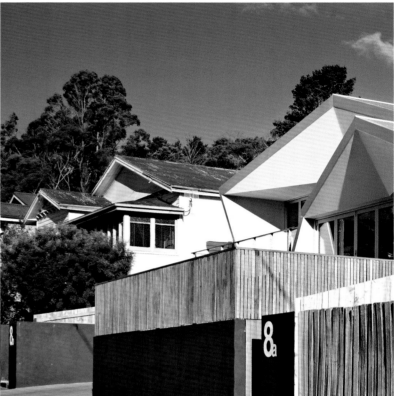
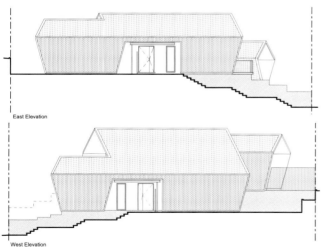

East Elevation

West Elevation

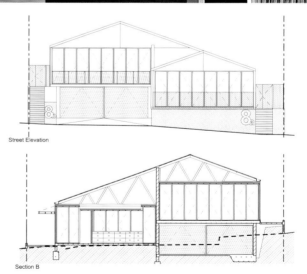

Street Elevation

Section B

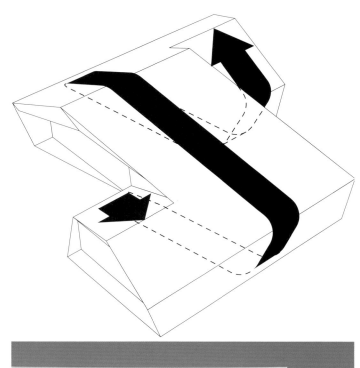
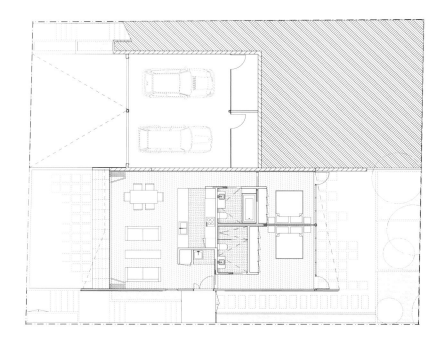
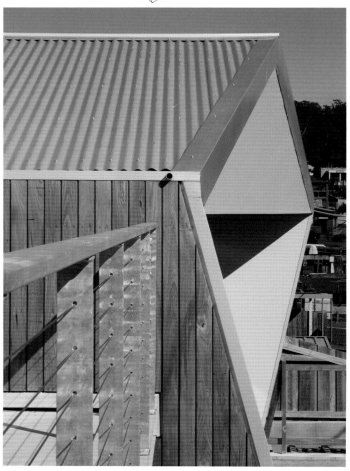

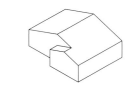
CALIFORNIA BUNGALOW REFERENCE TYPOLOGY

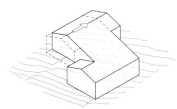
HORIZONTAL SHEER TO ADDRESS STREET CROSSOVER AND PROVIDE NORTHERN OPEN SPACES

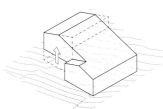
VERTICAL SHEER TO MINIMISE CUT & FILL

EXTENSION AND WRAP OF DOUBLE-GABLE ROOF FORM TO MAXIMISE PASSIVE SOLAR PERFORMANCE

THE LATEST VILLA APPEARANCE VI 069

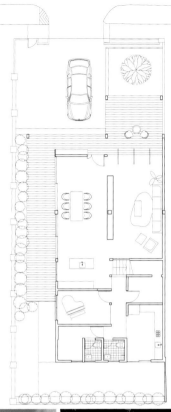

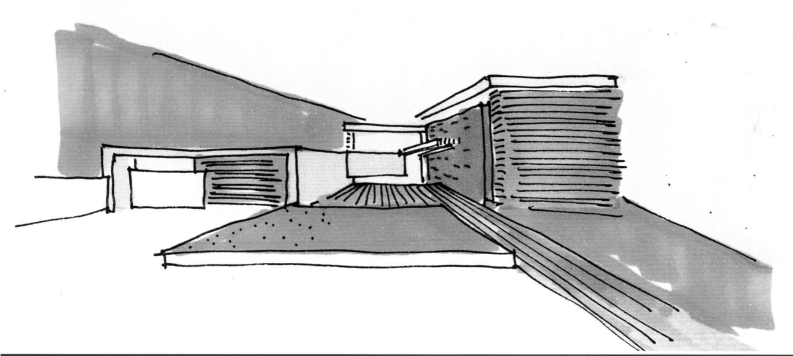

074 最新别墅外观 VI

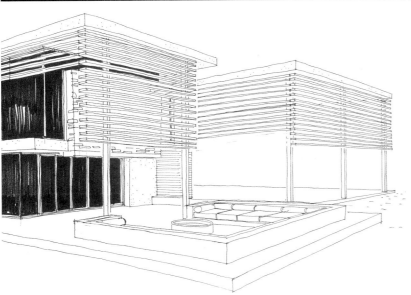
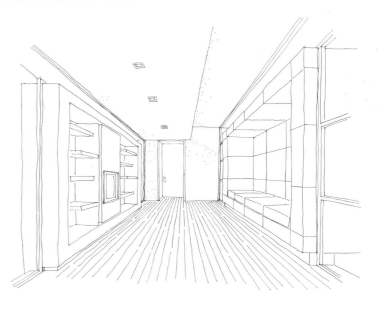

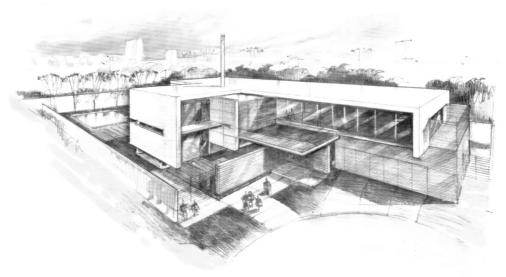
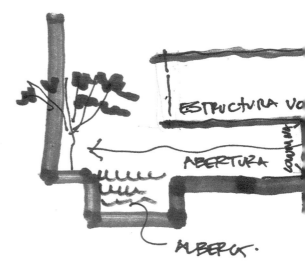

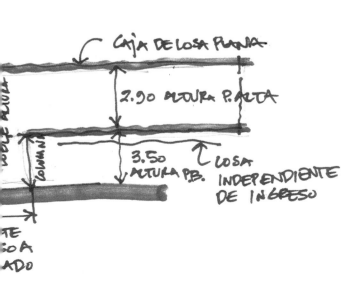

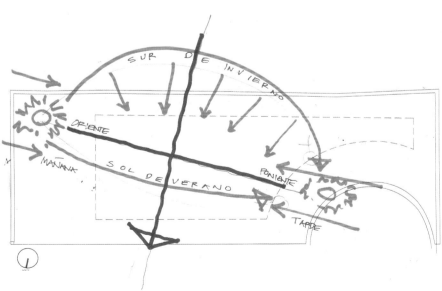

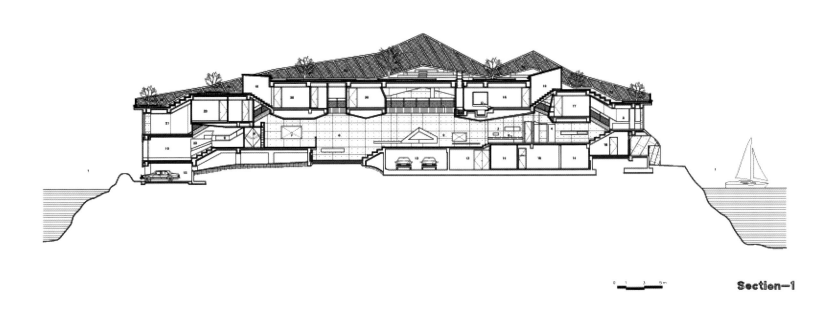

Section-1

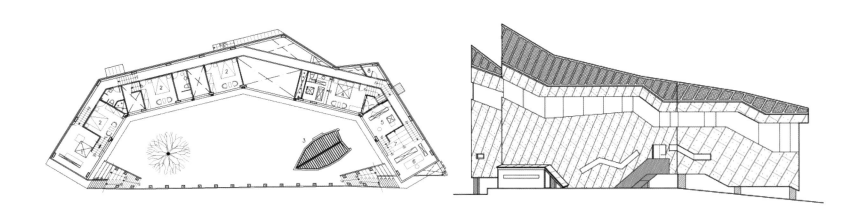

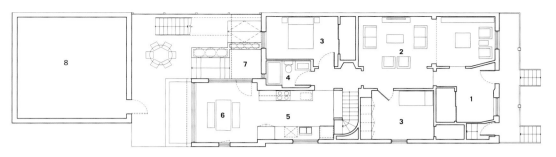

PLAN DE-REZ-DE CHAUSSEE

1. VESTIBULE EXISTANT
2. SALON EXISTANT
3. CHAMBRE EXISTANTE
4. SALLE DE BAIN EXISTANTE
5. NOUVELLE CUISINE
6. NOUVELLE SALLE A MANGER
7. NOUVELLE TERRASSE EXTERIEURE
8. GARAGE EXISTANT

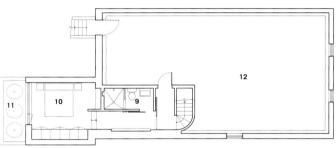

PLAN DU SOUS SOL

9. NOUVELLE SALLE DE BAIN
10. NOUVELLE CHAMBRE DES MAITRES
11. NOUVEAU JARDIN
12. SOUS-SOL EXISTANT

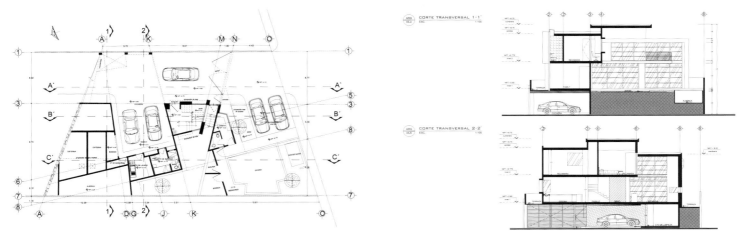

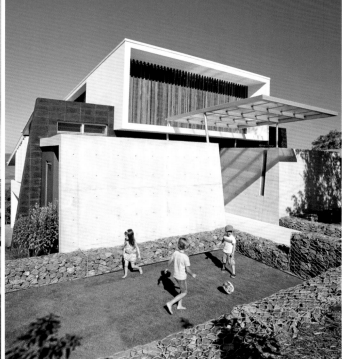

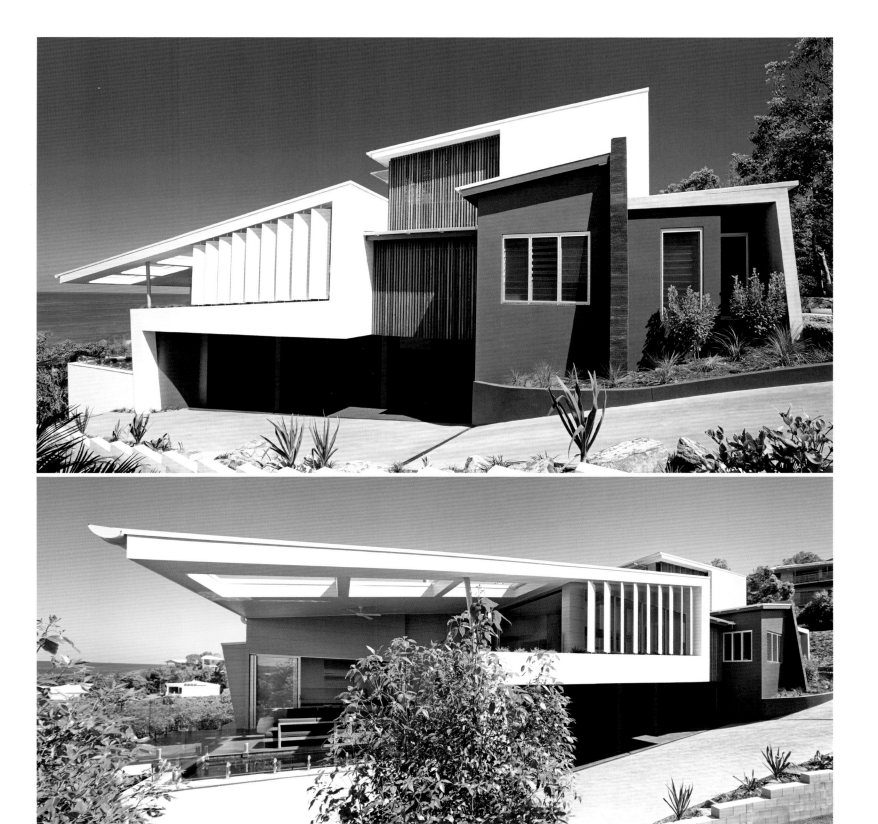

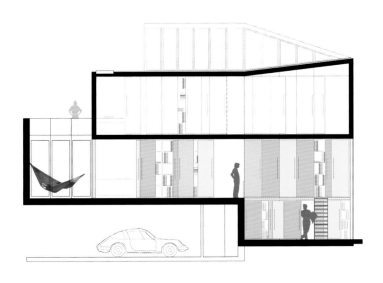

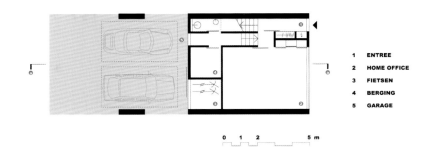

1	ENTREE
2	HOME OFFICE
3	FIETSEN
4	BERGING
5	GARAGE

0 1 2   5 m

1	WONEN
2	ETEN
3	KEUKEN
4	KAST
5	VIDE
6	PATIO

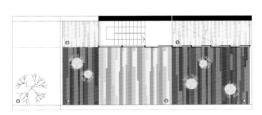

1	TUINKAMER
2	OPTIONEEL X-TRA KAMER
3	DAKTUIN
4	LIGWIJDE
5	DEK
6	VIDE

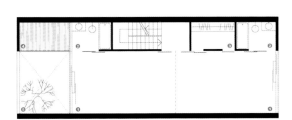

1	SLAPEN
2	BADKAMER
3	INLOOPKAST
4	TERRAS
5	VIDE

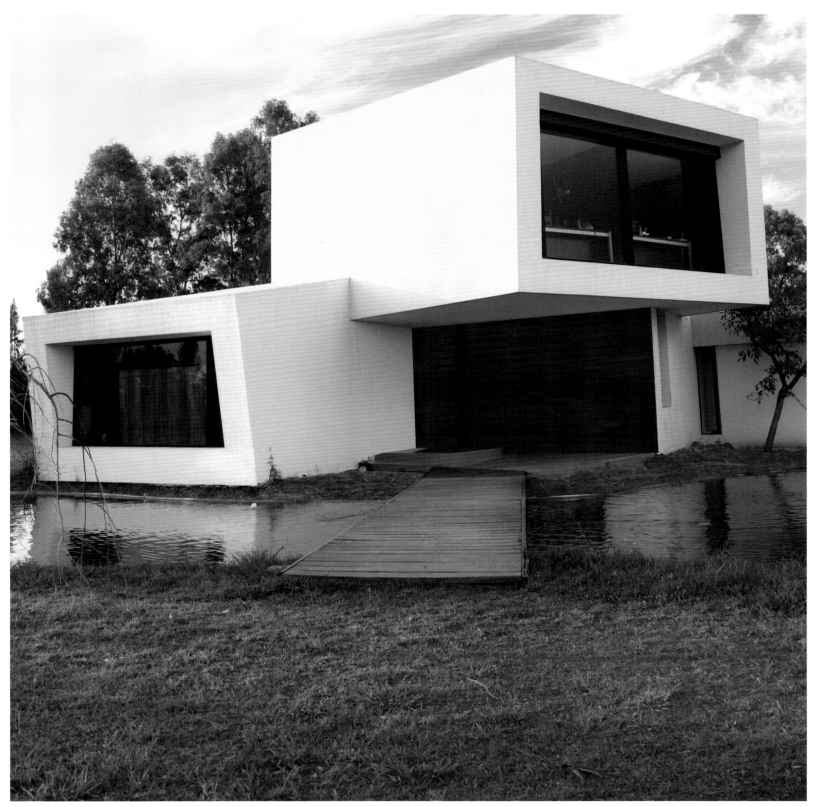

Various Style Villa

多样风格别墅

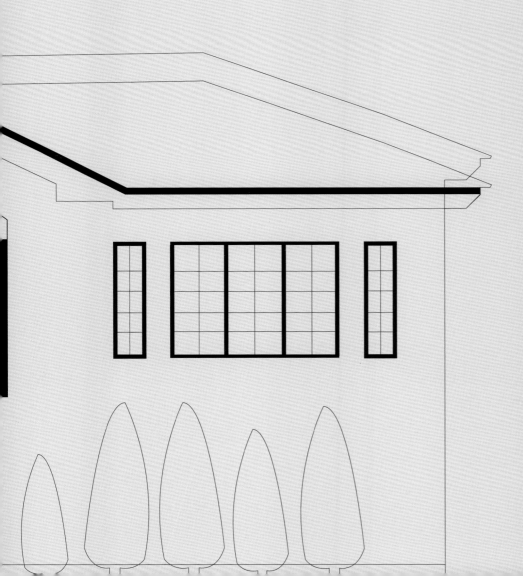

THE LATEST VILLA APPEARANCE VI

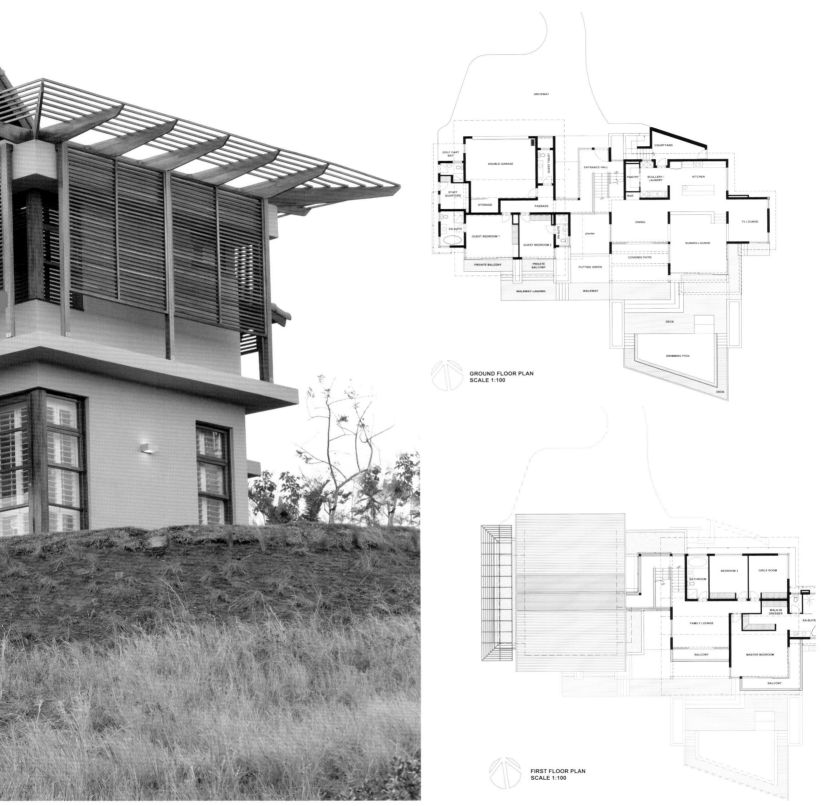

THE LATEST VILLA APPEARANCE VI

THE LATEST VILLA APPEARANCE VI

THE LATEST VILLA APPEARANCE VI

THE LATEST VILLA APPEARANCE VI

THE LATEST VILLA APPEARANCE VI

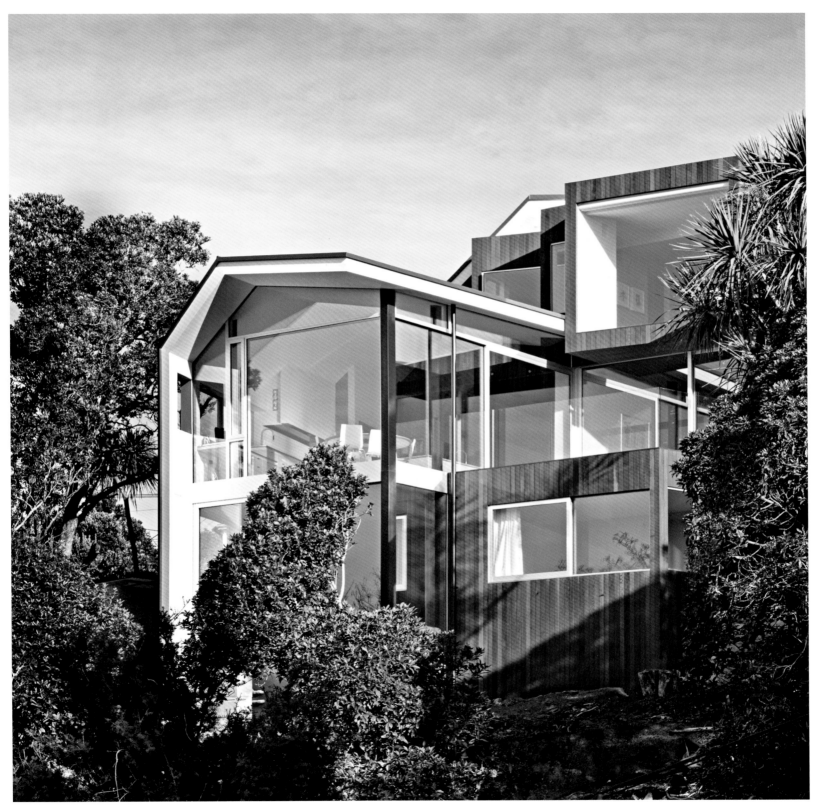

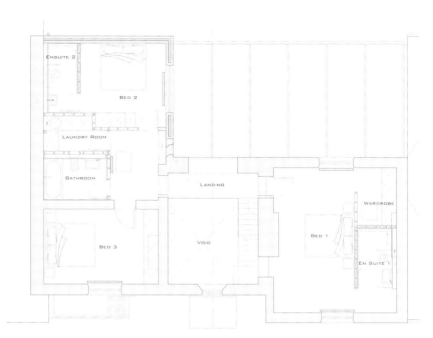

FIRST FLOOR PLAN

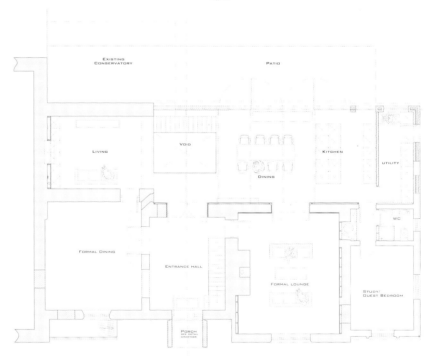

GROUND FLOOR PLAN

图书在版编目（CIP）数据

最新别墅外观Ⅵ / DAM 工作室 主编 . – 武汉：华中科技大学出版社，2013.7

ISBN 978-7-5609-9278-5

Ⅰ．①最… Ⅱ．① D… Ⅲ．①别墅 – 室外装饰 – 建筑设计 – 图集 Ⅳ．① TU241.1-64

中国版本图书馆 CIP 数据核字（2013）第 182337 号

最新别墅外观Ⅵ

DAM 工作室 主编

出版发行：华中科技大学出版社（中国·武汉）	
地　　址：武汉市武昌珞喻路1037号（邮编：430074）	
出 版 人：阮海洪	
责任编辑：熊纯	责任监印：张贵君
责任校对：王莎莎	装帧设计：筑美空间

印　　刷：利丰雅高印刷（深圳）有限公司
开　　本：1016 mm × 1058 mm　1/16
印　　张：15.75
字　　数：126千字
版　　次：2013年11月第1版 第1次印刷
定　　价：258.00元（USD 49.99）

投稿热线：（020）36218949　　　1275336759@qq.com
本书若有印装质量问题，请向出版社营销中心调换
全国免费服务热线：400-6679-118 竭诚为您服务
版权所有　侵权必究